Anne Caroline Lima da Silva Santos

Haitian Migration to Southwest Acre

Anne Caroline Lima da Silva Santos

Haitian Migration to Southwest Acre

Microregion of Brasiléia/AC

ScienciaScripts

Imprint

Any brand names and product names mentioned in this book are subject to trademark, brand or patent protection and are trademarks or registered trademarks of their respective holders. The use of brand names, product names, common names, trade names, product descriptions etc. even without a particular marking in this work is in no way to be construed to mean that such names may be regarded as unrestricted in respect of trademark and brand protection legislation and could thus be used by anyone.

Cover image: www.ingimage.com

This book is a translation from the original published under ISBN 978-613-9-68937-8.

Publisher:
Sciencia Scripts
is a trademark of
Dodo Books Indian Ocean Ltd. and OmniScriptum S.R.L publishing group

120 High Road, East Finchley, London, N2 9ED, United Kingdom
Str. Armeneasca 28/1, office 1, Chisinau MD-2012, Republic of Moldova, Europe
Printed at: see last page
ISBN: 978-620-8-24132-2

ACKNOWLEDGMENTS

First of all, I thank God for having sustained me this far. Then to my parents, family and especially my dear mother, Maria Luiza, who has always supported me lovingly on my journey. I would also like to thank Luan Marcel, who has always been by my side, encouraging me to be a better person in every way, both professionally and personally. I am also grateful to the dear friends I have made during my life and who have contributed in some way to this journey. Finally, I would like to thank the Haitian immigrants in Brazil, who gave me the chance to get to know their stories more closely and understand all the historical, environmental and social events that motivated this migratory process.

"The devaluation of the human world grows in direct proportion to the valuation of the world of things"
Karl Marx

SUMMARY

Migration is a broad and complex social phenomenon. It is therefore important to study the recent Haitian migratory movement to Brazil, which since December 2010 has used the territory of Acre as a gateway to the rest of the country, where they move to various Brazilian cities located mainly in the Southeast and South regions, particularly São Paulo, as well as other cities in Rio Grande do Sul and Santa Catarina. The Haitian migratory flow to Brazil is something new. The spatial mobility of populations has always occurred throughout human history, and the search to understand it finds the most appropriate method to achieve certain objectives, which is why we can consider the neoclassical and neomarxist approaches as the main paths of migratory research. This is why we want to understand why Haitians chose the illegal route to enter Brazil. And why did they choose Acre as their route, when it is located far from the country's major economic centers? Is it because of the support they received from the government? Ease of entry at the border? Easy access to documentation for legalization? A route without too many controls? This migratory flow began in 2010 after an earthquake measuring 7.0 on the Richter scale. According to the Haitian government, more than 223,000 people died and 1.5 million people were internally displaced.

Key words: Acre, Migration flow, Haitians, Mobility.

INTRODUCTION

The migratory process, whether national or international, brings with it many transformations. The country that loses its population as a result of the migration process suffers consequences that can be considered positive or negative, and when we think about the final destination, where these migrants will actually settle. We realize that the place will suffer a great influence in all aspects of the social sphere, be it cultural, political or economic.

The year 2010 marked the beginning of Haitian immigration to Brazil. The influx of Haitians began to be noticed in February 2010, shortly after the earthquake, which violently shook Haiti and the capital in particular. The catastrophe killed approximately 223,000 people and left 1.5 million people with nowhere to go. As a way of fleeing this natural disaster and the resulting lack of jobs, housing and new opportunities, these Haitians decided to come to Brazil. They started by entering the northern region of the country, through the states of Acre and Amazonas, where they could stay indefinitely, for a short time or as a transit route to other regions of the country.

From the outset, the earthquake became the Brazilian government's explanatory argument for Haitian migration to the country. So began the media visibility of Haitian migrants in the public debate on migration policies and citizenship processes of international migration in Brazil. As a result, the emergence of a solution to question Brazil's migration policy and the citizenship processes of Haitians in Brazil became paramount.

The work has three chapters: The first identifies and characterizes the municipalities of the Brasiléia micro-region, seeking to associate them with IBGE data. The second gives a brief history of Haiti, from its territorial formation to its social, economic and political characteristics. The third chapter deals with the natural tragedy that devastated Haiti and set the stage for the Haitian population to immigrate to Brazil in search of a fresh start.

The aim of this study is therefore to find explanatory factors for the recent

migration of Haitians to Brazil through the state of Acre, where the port of entry
is located.

CHAPTER 1

Theoretical and Conceptual Framework

1.1. METHOD OF APPROACH

The spatial mobility of populations has always occurred throughout human history. The search to understand it finds the most appropriate method to achieve certain objectives, which is why we can consider the neoclassical and neomarxist approaches as the main paths of migratory research. Until the 1970s, the phenomenon of migration was considered from a neoclassical perspective, within a predominantly descriptive and dualistic vision.

Neoclassical spatial mobility is characterized by the personal factors of migrants, the personal desire to migrate, disregarding social complexity. The neoclassical view studies migratory movements based on demographic flows and the individual characteristics of migrants. Each person sought to maximize their needs and the decision to migrate was seen only as a personal decision.

These theories disregarded the problem of urban unemployment and underemployment in poor countries, ignoring the proportion of the workforce that was not absorbed by the modern economy. Migration was seen as a balancing mechanism for changing economies, especially the poorest ones.

However, neo-Marxist spatial mobility approaches the historical-structural context from a dialectical perspective and considers the social process, understanding migration as a result of the introduction of capitalist production relations. From the mid-1970s onwards, the phenomenon of migration was reconsidered from a neo-Marxist perspective. Migration came to be seen as "mobility forced by the needs of capital" and no longer as a sovereign act of personal will.

The neo-Marxist approach to migration studies enables a broader understanding of reality. The vision of the constant and contradictory movement of society provides the necessary basis for migration research.

1.2. METHODOLOGICAL PROCEDURES

The starting point for the research will be a series of official documents collected from State Secretariats in Acre, SPF/AC (Federal Police Superintendence in Acre), DME/AC (Ministry of Labor and Employment Office in Acre) and possibly the Federal Revenue Office in Acre, as well as committees, councils, NGOs (Non-Governmental Organizations) and/or permanent national and international agencies, such as: CONARE/MJ (National Committee for Refugees - Ministry of Justice), CNIg/MTE (National Immigration Council - Ministry of Labor and Employment), UNHCR (United Nations Self-Commissioner for Refugees), as well as religious entities and dissertations, theses, also publications in books, magazines, newspapers, government plans, laws and statutes on immigration.

In addition to these sources, we also carried out the following steps: 1. Cabinet work; 2. Secondary data collection; 3. Analysis of data from various sources; 4. Thematic readings; 5. Writing the Capstone.

1.3. STUDY AREA - BRASILÉIA MICRO-REGION

The state of Acre is located in the extreme southwest of the Brazilian Amazon and corresponds to 4% of the Brazilian Amazon area and 1.9% of the national territory. The state has twenty-two municipalities and the capital is the city of Rio Branco, with a population of 733,559 inhabitants and a population density of 4.47 inhabitants/km^2 . It is divided into five micro-regions established by the IBGE.

. The population of the Brasiléia micro-region was estimated in 2016 by the IBGE at 66,106 inhabitants, the most populous municipality being Brasiléia with 24,311 inhabitants. As in the state as a whole, the municipalities have an average HDI, with Epitaciolândia having the best in the region. With the exception of Brasiléia, all the municipalities have fewer than twenty thousand inhabitants.

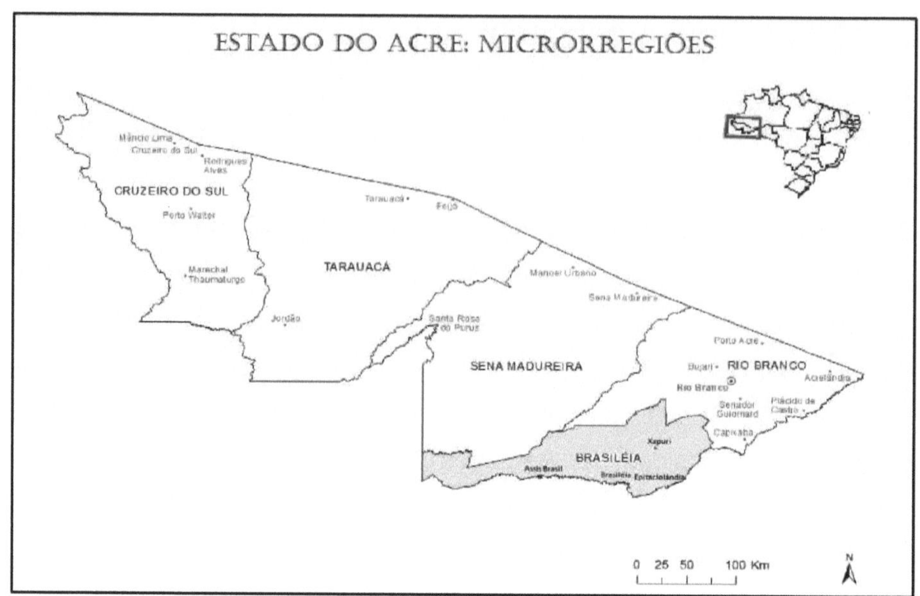

Figure 1: Map of the Brasiléia micro-region, where the municipalities of Assis Brasil, Brasiléia, Epitaciolândia and Xapuri are located. Source: IBGE 2016 - Modified by Anne Caroline 2017.

The GDP of the micro-region was estimated by the IBGE in 2015 at R$ 416,758. The municipality with the highest GDP is also Brasiléia with R$ 155,597 and has a total area of 3,916 km^2 , but the largest municipality in terms of area is Xapuri with 5,347 km2.

The cities in this region are characterized by being located in the Upper Acre River Basin and have very different characteristics, as shown in Table 1.

Municipality	Area Kmi	Population (2016)	Demographic density	GDP (IBGE, 2016)	GDP per capita (IBGE, 2016)	IDH (PNUD, 2010)
Assisi Brazil	**4.974**	6.863	1,22	54 375,889	10.985,35	0,588
Brasiléia	3.916	24.311	5,46	174.720,651	14 608,24	0,614
Epitaciolândia	**1.654**	17 038	9,13	127 947,833	15 596,52	0,653
Xapuri	5.347	17.894	3,01	129.949,450	11 385,49	0,599

Table 1: Socioeconomic data of the municipalities of Assis Brasil, Brasiléia, Epitaciolândia and Xapuri. Source: IBGE 2016 - Modified by Anne Caroline 2017.

1.4. CHARACTERISTICS OF THE MUNICIPALITIES IN THE MICRORREGION OF BRASILÉIA

The micro-region of Brasileia is a subdivision of the Meso-region of Vale do Acre, in the Brazilian state of Acre. It is divided into four municipalities: Assis Brasil, Brasiléia, Epitaciolândia and Xapuri.

1.4.1. ASSIS BRASIL

Assis Brasil is a Brazilian municipality in the state of Acre and is located on the triple border between Brazil, Peru and Bolivia, forming a conurbation of neighboring populations with the Peruvian city of Inapari and the Bolivian city of Bolpebra. The town is served by the BR-317 highway (Rodovia Transoceànica), which is the only highway linking Brazil to Peru.

1.4.2. BRAZILIA

Brasiléia is a Brazilian municipality located in the south of the state of Acre, 237 km south of Rio Branco, on the border with Bolivia, bordering the municipalities of Epitaciolândia, Assis Brasil, Sena Madureira and Xapuri. Despite being established as a free trade area, it has not yet been regulated. Currently, there is a strong trade dependency with the neighboring Bolivian municipality of Cobija. According to the IBGE (2016), the municipality ranks sixth in number of inhabitants, with 24,311 inhabitants, in the proportion of 64.22% urban, 15,612 inhabitants, and 35.78% rural, 8,699 inhabitants. Of these, 1,060 are riverside dwellers who live in communities on the banks of the River Acre.

1.4.3. EPITACIOLAND

Epitaciolândia is a Brazilian municipality in the interior of the state of Acre, bordered to the north by the municipality of Xapuri, to the south and east by Bolivia and to the west by the municipality of Brasiléia. Economic activity revolves around commerce, both for Brazilian residents and Bolivians, as well as livestock farming. It has several hotels, which serve as a base for Brazilians coming from Rio Branco to shop in the Cobija Free Trade Zone.

In fact, the economies of the three cities are complementary. Epitaciolândia stands out for having the largest bank in the region (Banco do Brasil) and its inhabitants look to Brasileia for postal and health services (Regional Hospital), as well as the advanced UFAC campus. Cobija stands out for its intense trade in the Free Trade Zone and for the University, which has also gained a lot of demand due to the offer of places on the medical course, which has led to many Brazilians coming from both the state of Acre and other states in Brazil. As a result, the local economy has grown even more.

1.4.4. XAPURI

Xapuri is a Brazilian municipality located in the interior of the state of Acre. It is considered a historic city, as it is the "cradle" of the Acre Revolution and the symbol of the global environmental movement. It is also known for the rubber tapper and union leader Chico Mendes, who lived all his life in the city and fought for many environmental and social causes.

Xapuri's economy is basically focused on the primary sector and livestock farming, with plant extraction standing out. Rubber and chestnuts are still the municipality's main products. The city is currently experiencing a trend towards the industrialization of forest products (rubber, nuts and wood). In 2008, the first natural condom factory on the planet, Natex, was set up in the city, which uses natural rubber from the region's extractive reserves to manufacture condoms. The municipality is also full of historical sites and monuments: The Church of São Sebastiâo, the Chico Mendes Museum and the Cachoeira rubber plantation, 40 minutes from the city center and home to the Chico Mendes Agro-Extractivist Settlement.

In this way, we can see that each municipality has its own particular history, economy and experiences. In this region of the Acre valley, there are memories of the Rubber Cycle and the struggle against Bolivian domination. The twin cities of Epitaciolândia and Brasiléia define the international border with Bolivia, as on the other side of the river is Cobija. Heading west is Assis Brasil, a town on the triple border with Bolivia and Peru, where the highway that integrates this region of Brazil with its South American neighbors and the Pacific Ocean passes through.

What we can highlight is the extensive border free of monitoring through Brasiléia, Eptaciolândia and Assis Brasil, which has facilitated the entry of Haitians into the country in search of opportunities such as work, medical assistance and the necessary aid that Haiti was no longer able to offer due to the natural disaster that occurred. After the earthquake, living conditions worsened, as did social and economic degradation.

CHAPTER 2

A BRIEF HISTORY OF HAITI

2.1. THE PROCESS OF TERRITORIAL FORMATION

Understanding the process of formation of the Haitian territory and its current precariousness consists of understanding the processes that have led Haiti to the sad situation of being among the poorest countries in the world and one of the consequences of this is the phenomenon of migration.

The history of the first independent Latin American country portrays an unstable political, economic and social trajectory right up to the present day. Revolts, coups and repressions have marked the Haitian people, who survive countless human rights violations. This is why the intention is to present the main events that reproduce the history of this country, highlighting their characteristics and importance for the formation of Haiti today.

2.2. - CHARACTERIZING THE TERRITORY

Haiti occupies the western third of the island of *Hispaniola*, with a surface area of around 27,750 km^2 . This corresponds to an area equivalent to the Brazilian state of Alagoas. To the north its territory is bathed by the Atlantic Ocean, to the south by the Caribbean Sea, to the west by the Bay of *Gonaïves* and to the east it shares the island of *Hispaniola* with the Dominican Republic, the only land border of which is about 360 km long. Haiti's territory consists mainly of rugged mountains with small coastal plains and river valleys (MOREL, 2017).

The climate is tropical and highly influenced by the sea. It is also located in the hurricane circuit, where there is a frequent incidence of tropical storms and hurricanes from June to October. The rainy season occurs twice a year from April to June and between October and November and these characteristics contribute to the history of frequent flooding.

The predominant vegetation is tropical rainforest, but it has suffered great devastation over time. Today, Haiti has very little vegetation for economic purposes and population concentration (MOREL, 2017).

The following map (figure 2) shows Haiti's territorial extension compared to the rest of the continent.

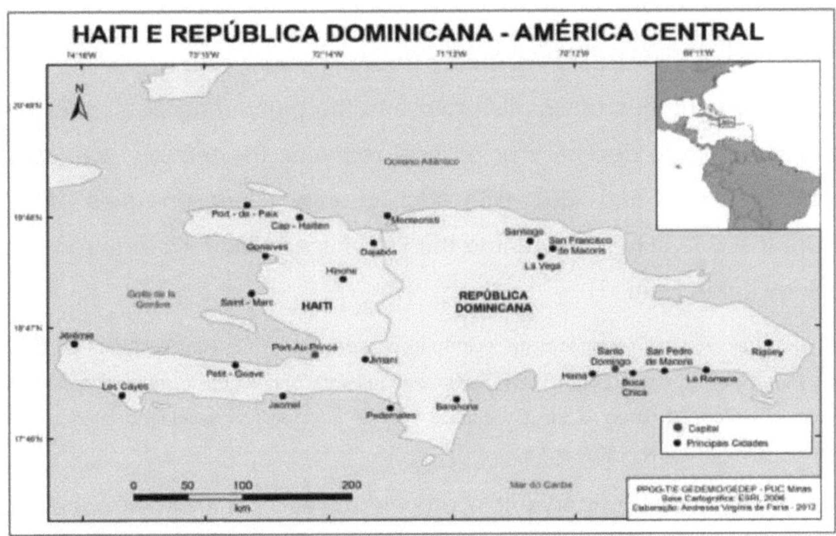

Figure 2: Map of Haiti and the Dominican Republic. Source: Faria (2012)

2.3. - A PAST FULL OF MARKS

The arrival of the European people in Haiti and Columbus' conquest took place in 1492, which ended up changing many things in the region. Christopher Columbus brought a lot of technology, cannons, soldiers and firearms, so the island was easily conquered with his armies eliminating almost all the Indians who didn't have the war technology to resist. After dominating the region, Christopher Columbus named the island *Hispaniola*, a name used to this day, possibly a tribute to his country Spain. (MOREL 2017)

Over time, domains of civilization are formed that absorb local environments, environments of civilization that impose a general norm that is imprinted on many ways of life. Islam, Hinduism and China represent types of superior civilization whose imitation extends far beyond the limits of their natural regions. The European plays the same role; the Yankee tends to take it over in America.

(LA BLACHE, 1954, p.377)

In this excerpt, La Blache makes us see that there are superior civilizations and they have always existed, imposing their norms and cultures on those civilizations that are new and considered lesser. This is exactly what happened in Haiti and in other states on the American continent, the imposition of a new general concept that everyone should follow.

After almost two centuries of Spanish rule and the plundering of all possible wealth, Spain began to decline. For political reasons, the territory was soon divided with France in mid-1697, with each country taking one side of the island. From the Spanish colony came the Dominican Republic and from the French colony came Haiti.

The French soon followed, with a successful colony in western Santo Domingo, which had been abandoned by the Spanish since 1605. Thus, Spain was left with only Cuba, Puerto Rico and the eastern region of Santo Domingo in the Caribbean - all of them still undeveloped islands with precarious colonization (KLEIN, 1987, p. 66).

Under French rule, Haiti underwent a process of economic growth and the exploitation of various products. Haiti became the most prosperous colony in the new world. Its fertile soil produced large harvests and attracted more French colonizers. Since the colonization period, Haiti has had a primary economy. It produced excellent quality sugar, which competed with Brazilian sugar in the 17th century and, together with all the production in the Antilles, helped to devalue Brazilian sugar in Europe. (MOREL 2017)

2.4. - IT WASN'T ALL PROSPERITY IN HAITI

With an economy heavily based on slavery, a wealthy French minority controlled an enslaved African majority with an iron fist. Haiti was made up of large farms and almost all non-agricultural items such as clothes, utensils, tools and weapons were brought from Europe, mainly France. While the ruling minority became richer and richer, the slave majority became more and more miserable. The inevitable revolt took place in 1794 when, after a violent uprising, Haitian slaves became the first country in the world to abolish slavery. (MOREL, 2017)

After this event, Haiti went through several rulers, starting in 1801 with Toussaint Louverture, a former slave and popular leader who proclaimed himself governor general of Haiti. However, a few months later, France sent continental military reinforcements and managed to defeat and assassinate the governor general. In 1804, in retaliation for the Haitian revolution, American and European slaveholders took France's side and set up a naval blockade against Haiti, which remained commercially isolated for 60 years.

After spending the entire 19th century in crisis and being administered by more than 20 rulers (of the 20, 16 were killed violently), in 1957, fearing that the communists in Cuba would turn Haiti into a republic allied with the Soviet Union, the Americans (it was the Cold War) helped the doctor François Duvalier (codenamed Papa Doc) to become president. Papa Doc established a hard-line dictatorship in Haiti, controlling the country by force and diverting a large part of the resources to corruption. Papa Doc managed to sink Haiti's already fragile economy even further (MOREL, 2017).

Pap Doc remained in power until his death in 1971, when he was replaced by his son Jean-Claude Duvalier (codenamed Baby Doc), who became Haiti's new dictator. Baby Doc continued in his father's line of corruption and persecution, facing several revolts during his rule. In 1986, unable to hold off the rebels any longer and fearing a new civil war, Baby Doc fled Haiti for France, taking his family and approximately 100 million dollars in embezzled money. (MOREL, 2017)

With Haiti sinking deeper and deeper into economic crisis and without external partners, the population has almost completely cut down the country's forests to produce and sell coal. This ecological disaster has changed Haiti's climate. Without most of its forests, Haiti has become drier, floods have increased and soil erosion has worsened. With a bankrupt economy and a devastated environmental area, armed conflicts spread throughout the country and chaos ensued.

2.5. - DEMOGRAPHIC DENSITIES OF THE HAITIAN TERRITORY

According to data organized by the UN (2014), in 2014 Haiti had a population of 10,461,409 people, 95% of whom are black, spread over 299.27 inhab. km^2 and its GDP in 2012 was a modest US$7.187 billion. With a relatively small area of land, Haiti has already achieved great proportions in the production of coffee and sugar.

The country's capital is *Port-au-Prince*, formed by the conurbation of four cities: the Arrondissement of Port-au-Prince and the communes of Delmas, Pétionville and Carrefour. The metropolis of Port-au-Prince is the country's main urban concentration, with around seven million inhabitants. Although Haiti's population density averages 270 inhabitants/km^2 , its population is concentrated in urban areas, coastal plains and valleys. Around 90% of Haitians are of African descent, the rest of the population is mainly mulatto of mixed Caucasian-African descent. A minority have European or Levantine blood and around two thirds of the population live in rural areas.

French is one of the two official languages, but is spoken by around 10% of the population. Almost all Haitians speak Krèyol (Creole), the country's other official language, which is a dialect derived from a mixture of African languages, French, English and Spanish, while English is increasingly spoken among young people and in the business sector.

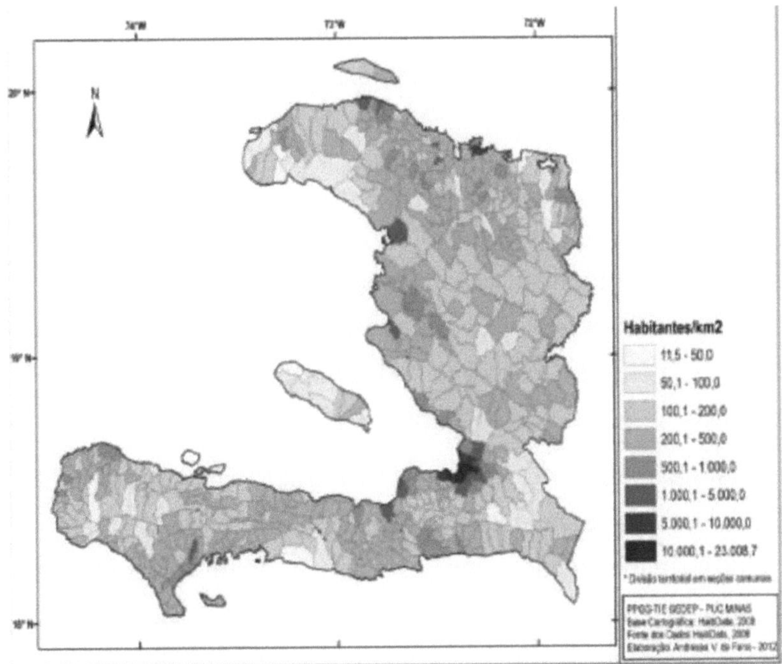

Figure 3: Haiti's demographic density - 2008. Source: Faria (2012)

2.6. social, economic and political characteristics of haiti

2.6.1. EDUCATION

According to data from the blog "SOS do aquecimento", the lack of formal education or its low quality is often one of the main roots of a country's social ills. The standard of education in Haiti is extremely low, where the literacy rate is around 54.8% (55% for men and 51% for women) and is well below the average of 90% of Latin American and Caribbean countries. Although education is free, school uniforms and materials are not, so many children are unable to attend school, especially in rural areas.

The best education is offered by private schools, most of which are run by Catholic or Protestant religious organizations. Some offer a complete education from kindergarten to secondary school. Other private schools are international and are run by countries such as Canada, France or the United

States. The country has a severe lack of school materials and qualified teachers, and the rural population continues to have little presence in the country's schools. (UN 2010)

2.6.2. HAITIAN WOMEN

As elsewhere, women's lives in Haiti are often very difficult. There they are responsible for maintaining the family and raising the children alone, suffer a lot of domestic violence and often sexual violence. In addition, birth rates are very high and so is maternal mortality, and access to health care and contraceptive methods is almost non-existent.

According to the data and accounts of Ana Maria Pereira, who currently lives in Port-au-Prince and works as a database administrator, she has written on a website called "brasileiraspelomundo.com" about some of the things she has experienced. Some of the research and data collected in Haiti indicates that around 40% of families are maintained by women, and they are also the ones who drive the food trade. In agriculture they also do most of the work: planting, harvesting and selling. Women are considered by many to be the backbone of the Haitian economy.

But unfortunately poverty often drives women into prostitution, especially when they don't have education, work, opportunities or a supportive family. The brutal reality for many women in this country is that they prostitute themselves on the streets of the poorest neighborhoods for 50 or 100 gourdes, which is equivalent to 1 or 2 dollars.

2.6.3. A WEAKENED ECONOMY

After several dictatorial regimes, today its main export product is still sugar, as well as other products such as bananas, mangoes, corn, sweet potatoes and vegetables. There is also livestock farming, which is simple with small herds of horses, cattle, goats and poultry. Another noteworthy activity is mining, which extracts marble, clay and limestone in small quantities. Fragile industry is concentrated in the areas of food (flour and sugar), textiles and cement.

However, this does not change the socio-economic problems experienced by the Haitian population, which has been marked by a series of dictatorial governments and coups d'état. Its economy is currently in ruins, and Haiti has become the poorest country in the Americas and the entire Western Hemisphere. Based on data obtained by *Trading Economics,* its human development index is 0.404 (low), approximately 60% of the population is undernourished and more than half lives below the poverty line, i.e. on less than 1.25 dollars a day. Haiti has an illiterate population of 45.2% and a life expectancy of just 60 years.

According to UN figures in 2003, 80% of the population live below the poverty line and more than half are destitute. A curious fact is that in order to reduce the hunger of the population, in 2004 Haitians were given a cake of butter, salt, water and a small portion of soil to satisfy their hunger. In places where poverty is really strong, many people end up eating small portions of land, a practice that comes from the ancient Indians.

Haiti's trade balance is in deficit according to 2009 figures obtained by *Trading Economics*: exports of US$ 558.7 million and imports of US$ 2.048 billion from the main trading partners: the United States - 33.11%, the Dominican Republic - 23.53%, the Netherlands Antilles - 10.75% and the People's Republic of China - 5.36%.

2.6.4. A DUBIOUS POLICY

Haiti's government is a semi-presidential republic, with a multi-party system in which the president is the head of state directly elected by popular vote. The prime minister acts as head of government and is appointed by the president, chosen from the majority party in the National Assembly. Executive power is exercised by the president and the prime minister, who together form the government. The constitution was introduced in 1987 and was modeled on the constitutions of the United States of America and France. It was partially or completely suspended for a few years, but returned to full validity in 1994.

Haitian politics have been controversial, as in its 200-year history Haiti has suffered 32 coups d'état. The country is the only one in the Western Hemisphere to have had a successful slave revolution, but a long history of oppression by dictators has significantly affected the nation. Countries like France, the United States and other Western countries have repeatedly intervened in local politics since the country's founding.

Together with the international financial institutions, they have imposed huge amounts of debt, so much so that foreign debt payments rival the government's budget for social spending. They have also applied economic policies that have undermined the Haitian government's ability to protect the local economy, forcing greater dependence on imports and undermining the country's economic self-sufficiency.

According to a 2006 Corruption Perception Index report, there is a strong correlation between political corruption and poverty, which puts Haiti in first place among all the countries surveyed for levels of internal corruption. The International Red Cross reported that seven out of ten Haitians live on less than two dollars a day. *Cité Soleil*, the country's largest slum in the capital Port-au-Prince (figure 4), has been called "the most dangerous place on earth" by the United Nations.

Figure 4: Cité Soleil slum.

Source: http://thecunninghamslife.blogspot.com.br/2010/11/where-jesula-grew-up-cite- soleil.html. Accessed on: 10/05/2017.

2.6.5. a country divided by departments

Haiti is territorially divided into ten departments: 1) Artibonite; 2) Centre; 3) Grand'Anse; 4) Nippes (created in 2003); 5) Nord; 6) Nord-Est; 7) Nord-Ouest; 8) Ouest; 9) Sud and 10) Sud-Est.

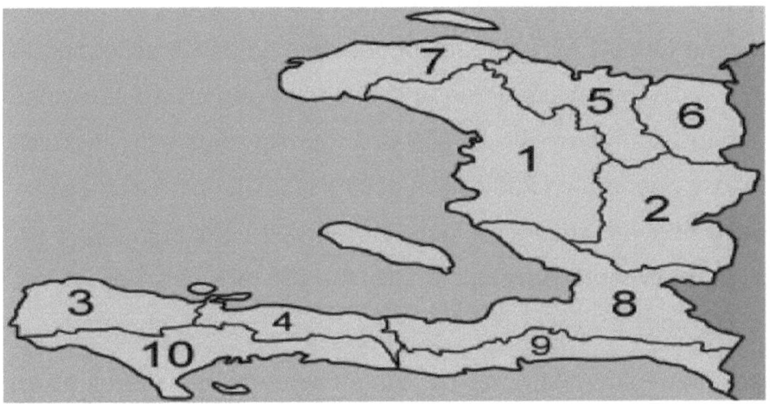

Figure 5: Haiti's territorial subdivisions.

Source: http://www.pcamaral.com.br. Accessed on: 12/05/2017.

Haiti is divided into the departments described in the table below:

Department Capital Area Population Density Map (Census 2003)

Artibcnite	Gonaives	4,895 km^2	1.070.397	218.67 inhabitants/km^2	1
Center	Hinche	3,597 km^2	565.043	157.09 inhabitants/km^2	2
Grande Cove	Jeremiah	3,100 km^2	603.894	n/a inhab/km^2	3
Nippes	Miragoane	n/a km^2	2.66.379	n/a inhab/km^2	4
North	Cape **Haitian**	2,175 km^2	773.546	355.65 inhabitants/km^2	5
North East	StrongFreedom	1,698 km^2	300.493	176.97 inhabitants/km^2	6
North West	Port-of-Peace	2,094 km^2	445.080	212.55 inhabitants/km^2	7
Of this	Porto Principe	4,595 km^2	3.093.699	673.28 inhabitants/km	8
South East	Jacmel	2,077 km^2	449.585	216.46 inhabitants/km^2	9
South	Les Cayes	2,602 km^2	627.311	241.09 inhabitants/km^2	10

Table 2: Subdivisions of Haiti.

Source: 2003 Census - Modified by Anne Caroline 2017.

Haiti is politically divided into departments and not into states as is the case here in Brazil. These Haitian departments are the first-level administrative divisions in the Republic of Haiti.

2.6.6. LOCAL TOURISM AND ITS CULTURES

Although Haiti is facing numerous economic and social crises, it still attracts the attention of some tourists for its beautiful beaches. In 2012, according to data obtained by the Terra website, in an article highlighting cruise itineraries to Haiti (figure 6), the country received 950,000 tourists, mostly from cruise ships, and this industry generated 200 million dollars in 2012. In the same year, several hotels were opened, such as a four-star Marriott hotel in the Port-au-Prince area and other new hotel developments in the capital and in Les Cayes, Cap-Haitien and Jacmel.

But in December of the same year, the US State Department issued a travel alert for the country, noting that although thousands of US citizens safely visit Haiti each year, foreign tourists have been victims of violent crime including murder and kidnapping, predominantly in the Port-au-Prince region.

Figure 6: Labadee, a cruise destination in the country.

Source: L. W. Yang, Los Angeles, USA 2007

Haitian Carnival has also become one of the most popular in the Caribbean since the government decided to stage the event in a different city each year and this has attracted more tourists. When we talk about Haitian culture, we see how it is a mixture of mainly French, African and Taino elements, with the influence of the Spanish from the colonial period. The country's customs are essentially a mixture of cultural beliefs that come from the various ethnic groups that have inhabited the island of *Hispaniola* over time.

Art is made up of bright colors, naive perspective and astute humor. Food and lush landscapes are favorite themes in this poorer land. Jungle animals, rituals, dances and the gods evoke the African past and are also present in the art. Haiti is known for its rich folk traditions.

23

Figure 7: Representation of a Haitian cultural festival

Source: http://haiticultura.blogspot.com.br/. Accessed on: 20/05/2017.

CHAPTER 3

A NATURAL TRAGEDY

3.1. THE GREAT EARTHQUAKE

From being the richest colony in the world in the 17th century to the poorest country in the Western Hemisphere, Haiti has spent the last 200 years suffering from military coups, violence, corruption, famine and natural disasters. The earthquake that practically destroyed the capital Port-au-Prince on January 12, 2010 was the worst tragedy in its history.

The United Nations (UN 2011) reported that more than half of the buildings in Port-au-Prince were destroyed, including the presidential palace. The city's infrastructure, which was already precarious, has been compromised, hampering humanitarian aid services and help for the injured, which is why the World Health Organization (WHO) has issued a warning about the risk of epidemics such as hepatitis A, tuberculosis and meningitis. The country is one of the poorest in the world, with 80% of the population living below the poverty line. It also has record levels of infant mortality, malnutrition and AIDS infection.

Another aspect that strikes us is the large concentration of people in the capital, Port-au-Prince, and its ability to polarize and interconnect the rest of the country. It was no coincidence that in the devastating earthquake of 2010, the largest concentration of deaths was in the capital.

At the departmental level, there is a predominance in the West department, where the presence of the metropolitan region of Port-au-Prince is an important element in the spatial distribution of the population, with all its economic, political and social consequences. In 2003, this department absorbed 23% of the country's population and 56% of the total urban population (IHSI, 2009 apud FARIA, 2012, p. 64).

On top of all these factors, there is also a long history of natural disasters that only worsen the situation of the country's inhabitants. Shortly before the 2010 earthquake, in 2008 more than a thousand people died and 800,000 were left homeless due to hurricanes that devastated the region, causing US$1 billion in damage, according to information obtained by the correspondent from international agencies.

Since its colonization, Haiti has suffered from a series of devastating natural phenomena that have been systematically aggravated by human action and also by the country's high population density. According to UN data (2005), only 0.11% of Haitian territory is protected. This figure is extremely low and helps to prove the country's neglect of environmental issues.

That's why we can say that the 2010 earthquake and its devastation were also exacerbated by the environmental impacts caused by man, as well as the precarious constructions that are the result of inequalities in the country. However, Haiti is located in a region of great tectonic instability, which naturally produces earthquakes and other natural phenomena.

The island of Hispaniola is located at a point of contact between the North American and Caribbean (Caribbean) Plates. While the North American Plate moves in an east-west direction, the Caribbean Plate moves in the opposite direction, meaning that this transcurrent movement makes the Caribbean a region of high seismic activity (FARIA, 2012, p. 74).

The earthquake that hit the country at 4:53 pm (local time) registered 7 on the Richter scale. According to the United States Geological Survey, the earthquake occurred at a depth of about 10 km and 22 km from Port-au-Prince. This first earthquake preceded two others of magnitudes 5.9 and 5.5, which contributed to increasing the damage in the cities. The earthquakes were caused by the movement of the Caribbean and North American tectonic plates, and Haiti lies exactly on one of the faults (the space between the two plates).

Earthquakes of this magnitude would cause damage in any country, but the historical conditions that make Haiti a nation lacking in almost all social support contributed to making the catastrophe worse. After this earthquake, during the last months of the same year, there was no real recovery process that could allow the people of Haiti to recover what existed before the earthquake. Haiti, whether in social or environmental terms, is becoming increasingly precarious, bringing instability to its inhabitants.

This event worsened job opportunities in the country, the formal education system, political life, food security, hygiene and health conditions, as well as further weakening the country's economy. The map below shows the location of Haitian territory in relation to the tectonic plate boundaries (Figure 8).

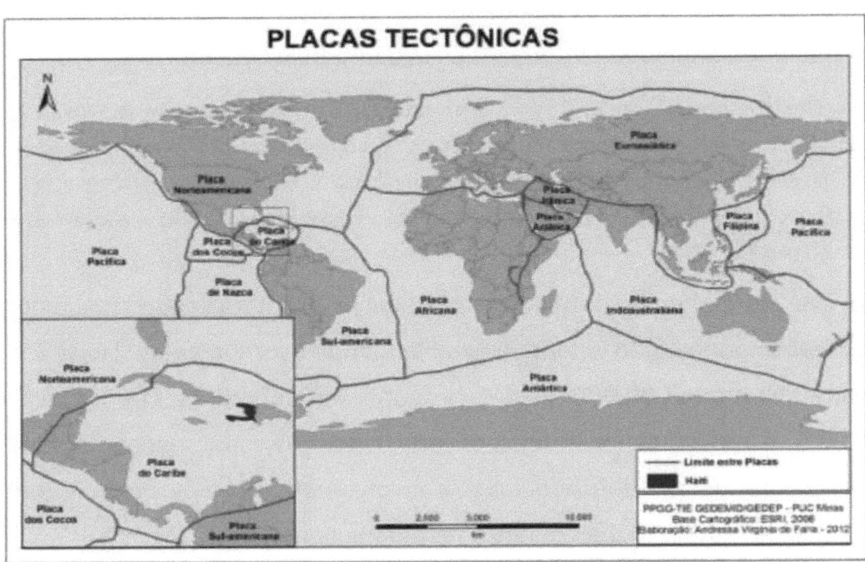

Figure 8: Location of Haiti and the tectonic plates. Source: Faria (2012)

As a result of the earthquake, some Brazilians were injured and killed. Brazil was responsible for the peace process in Haiti, commanded more than 7,000 United Nations (UN) peacekeepers and had 1,266 military personnel in the country. A week after the earthquake, 21 Brazilian deaths were confirmed, 18 of them military personnel and three civilians. Among them was the doctor Zilda Arns Neumann, who was the international coordinator of the Pastoral da Criança.

3.2. THE MIRAGE OF HAITI'S RECONSTRUCTION

Rebuilding a country after a natural disaster is a tireless process and the results are not always so visible. Even a few years after the earthquake that shook Haiti, in 2015 there were still civilians, the military and international aid working hard to get the country back on its feet.

In Haiti, 100,000 houses were completely destroyed and more than 200,000 seriously damaged, leaving more than one and a half million people homeless. Rebuilding housing seemed to be an urgent priority, but despite progress, the task has not been successfully completed. Figures published by *Forum Magazine* show that only 1% of emergency aid and 16% of reconstruction support has been channeled through the Haitian administration.

Five years after the tragedy, an EBC news team returned to Haiti to see how the country's reconstruction efforts are going. There are remnants of the quake in several cities. Notre Dame Cathedral, the main cathedral in Port-au-Prince, is still in ruins. You can only see the rubble of what was once the main church in the Haitian capital. The Haitian government building is also awaiting reconstruction (Aguiar, 2015).

In 2015, five years after the earthquake, UN figures (2015) estimated that more than 1.4 million people had left the makeshift camps over the years. However, there were still around 80,000 displaced people throughout the country, with around 80% of the population living in poverty. Unemployment reached around 30% of Haitians despite the population's involvement in many projects that generated work related to the country's reconstruction. (UN 2015)

It is estimated that more than 2 million Haitians have left the country because of political turmoil, better job offers or living conditions. After the earthquake, it was no different: thousands of Haitians crossed borders to restart their lives in other countries, including Brazil. At least 30,000 of them have arrived in Brazil since 2010, carrying memories of a day that changed their lives.

3.3. BRAZIL AS A "SOLUTION"?

According to the Ministry of Labor, in 2010 Brazil's economy created 2.52 million formal jobs with a formal contract, a new record in the historical series of Caged (General Register of Employed and Unemployed), which began in 1992. The creation of formal jobs broke records in 6 of the 12 months of 2010, and the number was the highest in history for the months of January to May and August 2010. (Caged/2010)

The sectors that showed an increase in employment were commerce and public utility industrial services (Caged/2010). The survey by geographic level reveals an expansion in employment in all the major regions and units of the Federation.

The Southeast leads with 1,276,903 jobs and a record in three states, followed by the Northeast with 488,561 jobs and a record in eight of the nine states and in third place the South with 444,713 jobs and a record in all states. Last was the Midwest with 178,242 jobs and a record in one state and the North with 136,259 jobs and a record in one state. (Caged/2010)

The sectors that hired the most were services with 500,177, commerce with 297,157 and construction with 177,185. The states with a positive balance in job creation were São Paulo with 277,573 jobs, Minas Gerais with 90,608 and Rio de Janeiro with 88,875. Between 2009 and 2011, the sectors that grew the most were Services, Commerce and Construction. The growth of the job market in Brazil at this time was also driven by programs such as PAC 2; Minha Casa Minha Vida, the 2016 Olympics and the 2014 World Cup. (Caged/2010)

So at this point we can see that for the Haitian nation, which had just been through a natural devastation and had no hope of new opportunities, and with a history of socio-economic struggles and losses, Brazil was seen as a solution for the Haitians. So they took courage and faced all the challenges and dangers to make it to Brazil in search of a fresh start.

3.4. MIGRATORY ROUTES USED BY HAITIANS

Many cultural, spatial, social and economic transformations took place during the journey to Brazil and during the journey through the interior of Brazil to their final destination. This journey was often marked by difficulties, upheavals and unforeseen events. The Haitians were often subjected to people who would "cross" them and who charged very high prices to bring them to Brazil, often in a completely degrading situation.

Haitian immigrants arrive in Manaus with a journey history of three months, never less than one month. A real ordeal. A story marked by anguish and suffering, by deprivation and hunger, of nights spent sleeping rough in tiny, cramped "lodgings". They arrive in Manaus exhausted and, for the most part, without a penny in their pockets (COSTA, 2011, p.84).

According to studies, the majority of Haitians interviewed said that it takes an average of US$ 3,000.00 and up to US$ 6,000.00 to get to Brazil. This amount in dollars is extremely high by Haitian standards and the sum is obtained through loans from family members or indebtedness to loan sharks, so the road traveled by immigrants is already difficult right from the start. (Cotinguiba/2014)

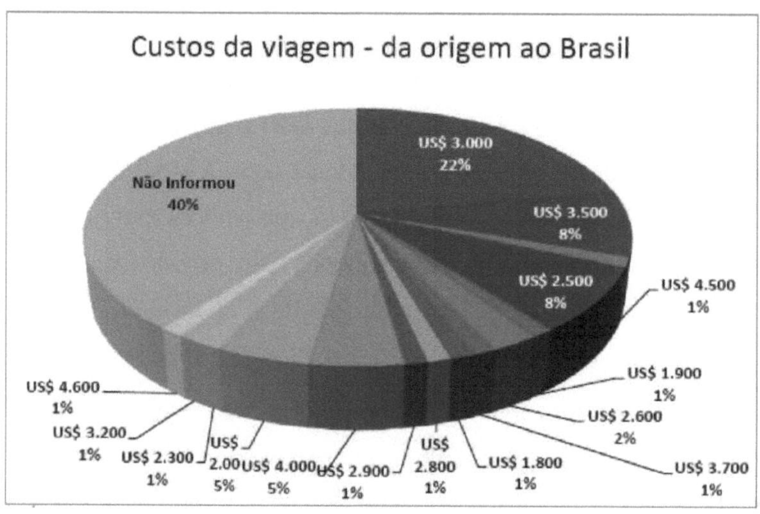

Graph 1: Travel costs from origin to Brazil. - Total interviewees, 173.

Source: Cotinguiba (2014)

This started a migratory flow that intensified in 2011 on Brazil's borders with Bolivia and Peru, through the cities of Brasiléia and Assis Brasil, in the state of Acre, and in Amazonas through the city of Tabatinga (Figure 9). The main route taken by Haitians to enter Brazil comprises a common point up to a certain part of the journey and then two for entry into the country.

The two entry points are Tabatinga, in the state of Amazonas, on the triple border between Brazil, Peru and Colombia, and the second, at another similar point, between Brazil, Peru and Bolivia. Of course, other routes can be taken, such as through Bolivia and Paraguay or even by plane to São Paulo, but this was not the most common way.

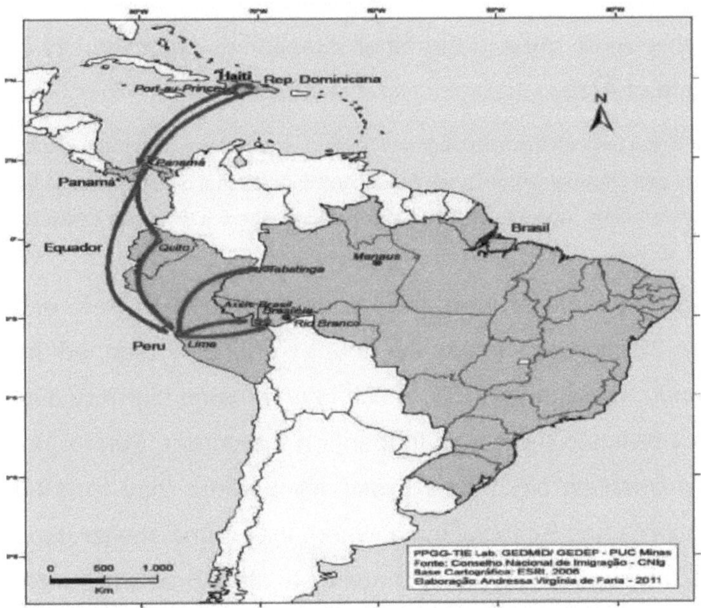

Figure 9: Migratory route used by Haitians to enter Brazilian soil.

Source:http://ponto.outraspalavras.net/2012/01/20/brasil-os-desafios-da-lei-de-migracoes/.

Accessed on: 10/06/2017.

You can see from the map the long way they have traveled to reach Brazilian territory, but between setting foot on Brazilian soil and arriving at their destination in a country as large as Brazil is a tiring journey, just like the journey from Haiti to the Brazilian border. There are various accounts of the difficulties faced by Haitians, ranging from hunger, lack of the minimum conditions of hygiene and comfort, to situations involving violence due to the long journey.

However, the main focus of our investigation is on the micro-region of Brasiléia, for the reason that this place of entry is linked to the flow of Haitians in other regions of Brazil. By January 2014, Brasiléia had received more than 18,000 Haitians, while Tabatinga in the state of Amazonas received less than 10,000. Some of these migrants left Haiti from different cities with a stopover in Panama

31

City, from where they went to Quito, the capital of Ecuador, and from there to Lima in Peru. At this point the route forks into two flows: for those heading for Amazonas and the other route goes in the other direction to enter the city of Assis Brasil, in the state of Acre.

If they were unable to enter Brazil through the front door due to legal requirements, the route initially planned by the agents was to pass through some South American countries that didn't require a visa to reach the Brazilian border, either in the state of Amazonas or Acre, where it would be easier for them to enter (SILVA, 2012, p. 304).

To enter Brazil through the state of Acre, the Haitians left Lima in Peru and followed the Carretera 3S highway across the country until they reached the city of Puerto Maldonado, on highway 26B, which later became highway 30C and through there they reached the city of Inapari, on the border. Further on, they reached the first Brazilian city: Assis Brasil, from where they traveled another 110 km to Brasiléia and Epitaciolândia, where they found shelter, food and the necessary documentation to try to continue on to other Brazilian states, passing through the capital Rio Branco.

In Brasiléia, the immigrants apply for refuge with the Federal Police and wait for their documents to be processed in unsanitary conditions. The shelter used to be inside an old club with a capacity for 250 people. In April 2013, around 1,300 immigrants were sleeping there, prompting the governor of Acre to declare a social emergency in the city.

Figure 10: Haitians looking for documents at the Federal Police station. Photo: Gleilson Miranda (2012)

Food for the newly arrived Haitians in Brasiléia depended on donations, either from the population or from the state government. Due to the increase in the influx of immigrants, the donations were often not enough to feed everyone. So the state government only started providing food for the immigrants again in February 2013, after hiring a company that sold lunch boxes at a very low price.

In April 2013, the Acre government opened a National Employment Service (SINE) office in Brasiléia. Support from the Ministry of Labor began in May, with the aim of speeding up the issuing of work cards. This registration in the employment system is not only in the interest of the Haitians, but also of businesspeople who have their eye on the labor force that arrives in droves across the borders. With no money in their pockets and the desire to find a job, they try to leave Brasiléia. Some rely on family for financial support, others work odd jobs or wait to be hired by a company that will provide them with airfare.

In December 2011, the Brazilian government announced the possibility of closing the borders in the region where Haitians were entering, or controlling the number of entries, and the result was hundreds of people entering in just a few days. Upon entering Brazilian soil, in the case of those going to Brasiléia, the Haitians boarded a taxi. These taxis charge between twenty-five and fifty dollars per person to take them to the refuge.

Some of the places used as shelters for the Haitians were the Nossa Senhora das Dores Parish. The second was the Sports Gymnasium in Epitaciolândia, and the third was the Hotel Brasiléia located in Hugo Poli Square, which ended up attracting attention due to the conditions present, as it was a place with a capacity for eighty people and at the time had more than a thousand people. The number increased every day and the long wait was up to three months. The delay was due to obtaining documents as a measure of detention for screening those entering the country.

The result of this measure was a violation of human rights, with people sleeping on the benches in the square, taking turns on the few mattresses there were or sleeping on cardboard (Figure 11). Another place that served as a shelter was a building belonging to a club, and with the exception of the Sports Gym, all the others are in Brasiléia. In April 2014, the shelter in Brasileia was closed and the activities of receiving immigrants transferred to Rio Branco. In January 2015, Acre's Secretary of Justice and Human Rights, Nilson Mourâo, asked for federal aid to assist Haitian refugees in the municipality of Brasiléia (EBC, 2015).

Figure 11: Haitians crowded into the shelter in Brasiléia. Photo: Machado (2014)

Figure 12: Place that served as a shelter for Haitians during their stay in Brasiléia. Today it is deactivated. Photo: Galvâo (2016)

According to Brazilian law, refugees can only be granted to those who can prove that they are suffering persecution in their country for ethnic, religious or political reasons. However, due to the humanitarian crisis caused by the catastrophe in 2010, the Brazilian government made an exception, granting them a differentiated visa (BAENINGER, 2013). It is also known that there are still many Haitians in illegal conditions due to bureaucratic factors and requirements made by the Ministry of Foreign Affairs for foreigners to enter the country.

After a lot of struggle and suffering, the Haitians finally achieved their first great victory, which was to leave Acre. Some went with a guaranteed job, others tried to find work through friends and relatives who had been in the country longer. At that moment, Rio Branco airport was filled with Haitians, especially in the early hours of the morning, when flights are cheaper.

Of the more than 30,000 Haitians who have crossed the Brazilian border in recent years, according to data from the government of Acre (2014), the main access to Brazil has been in São Paulo or in the southern states, mainly Rio Grande do Sul and Santa Catarina. For Haitians, Acre has become a gateway and a place of passage.

3.5. PROFILE OF HAITIAN IMMIGRANTS ARRIVING IN BRAZIL

The three parameters that will be analyzed, age, gender and schooling, form an important basis for better understanding the characteristics of these Haitian immigrants.

According to John Wilmoth (2013), Director of the Population Division of the United Nations Department of Economic and Social Affairs (UN-DESA), 74% of international migrants are of working age, i.e. between 20 and 64 years old, and 48% of migrants are women. This general data provided by the UN-DESA (2013) shows the intensification of migration in recent decades, especially the migration of workers, and that this trend also applies to the Haitian case.

According to various sources and reports from migrant pastoral organizations dealing with this issue, the total number of Haitian immigrants living in the country is between 18 and 34 years old. One example is the information provided by Justice Minister José Eduardo Cardoso on April 18, 2013, when he was in the city of Brasiléia in the state of Acre and indicated that there were 1,300 immigrants in the city (including 815 Haitians) and that the majority were aged between 18 and 25. The age group of Haitians arriving in Brazil, especially since 2011, shows us the nature of the spatial mobility of the workforce, since the vast majority of these immigrants were in full physical and intellectual condition to carry out some kind of job in our country, especially jobs related to physical strength.

The gender of Haitian immigrants in Brazil is another issue worth considering. According to data from the National Immigration Council (CNIg) and the Ministry of Labor and Employment (MTe), of the visas granted to Haitians in 2011 and 2012, which totaled 5580 people, 87.1% were men.

Visas granted to Haitians (2011-2012)

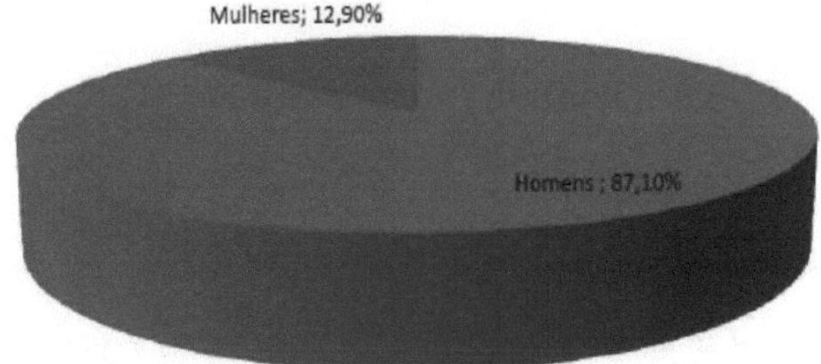

Graph 2 - Visas granted to Haitians between 2011 and 2012

Source: Ministry of Labor and Employment, National Immigration Council. Fernandes, Castro (2014)

The graph shows the great disparity between male and female immigrants. The trend is certainly towards an increase in Haitians in Brazil and, consequently, an increase in the number of women will also be noticeable, since with a few years of presence in our country and with more structure, they will be able to bring their families and friends who stayed in Haiti.

When we look at the age distribution between 2010 and 2014 (graph 3), we see that there is an increase in the child population, which also suggests this process of family reunion, as previously mentioned due to the better structure achieved or also due to the arrival of women who were already pregnant and had faced a lot of difficulty until the birth of their children.

Following the characteristic of international immigrants, whose main objective is to find a job that will build something productive and generate money, Haitian immigrants are relatively young.

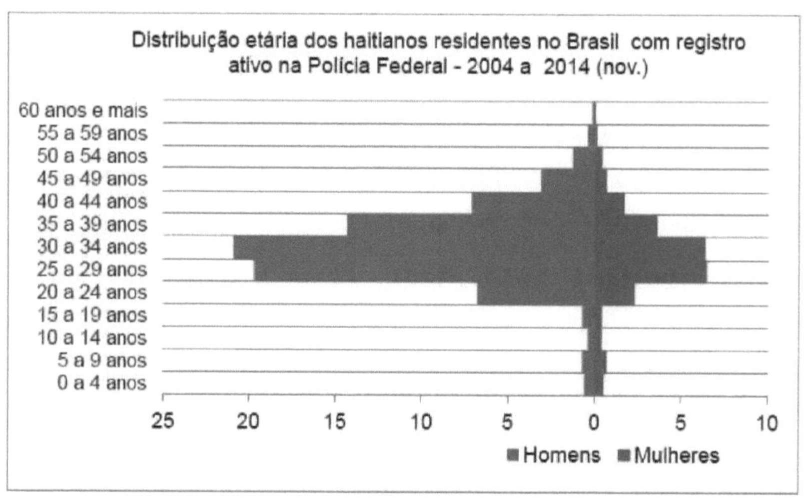

Graph 3: Age distribution of Haitians living in Brazil. Source: Federal Police Register (2004 - 2014)

Now, when we look at the forms of school access and organization in Haiti, we see that they are very different from those found in Brazil. These data will only serve as an indication of this.

With regard to education levels, it is important to remember that the Haitian school system is different from the one in use in Brazil, especially in terms of grades and levels of education. Thus, the data on education contained in the CNIg databases should be considered as indicative of the level of education and possibly does not correspond to the actual schooling of immigrants. (FERNANDES, CASTRO, 2014, p.29)

Companies have realized that even without a minimum level of education, without taking into account the differences in the school structure between the two countries and knowledge of the language, it is still good business to hire this workforce, which often does the same job as a Brazilian worker for a lower wage, and does not receive the same benefits.

However, another aspect of this story is that a number of Haitian immigrants in Brazil have completed higher education or have studied in higher education but not completed it, in professions such as engineers, doctors and journalists.

The profile of the immigrants seeking places in the church is very diverse. There are Haitians with no schooling and Nigerians who arrived in the country illegally, hiding in the holds of ships with only their clothes on. But there are also journalists, doctors and engineers among those coming from Haiti (SANCHES, 2014).

So we can see that there are two types of Haitians: those with almost no basic education and those with very important and even essential training, such as doctors. Some data provided by the MTe, together with the CNIg, gives us an idea of the level of education of Haitians.

Level of education of Haitians applying for residence permits in 2012

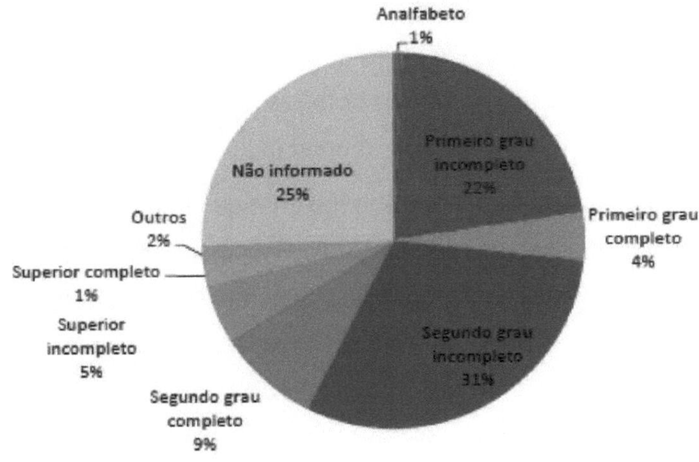

Graph 4: Level of education of Haitians.

We can see from the analysis of the graph that the majority of Haitians who migrated to Brazil in 2012 had incomplete secondary education. Another statistic that stands out is that 25% of Haitians did not want to or were unable to state their level of education.

3.6. HAITIANS IN THE BRAZILIAN LABOR MARKET

The search for work in Brazil has undoubtedly become the main objective of Haitian immigrants. This created a relationship between immigrant workers looking for better salaries in order to change their living conditions and thus be able to send money to their families who remained in Haiti and, on the other side, the companies who saw these workers as a chance to increase their profits by exploiting them more, thus extracting surplus value. Many Haitians reported on the construction of the Belo Monte power plant, which was going to hire 25,000 workers in one go, which aroused even more interest among Haitians in Brazil.

The Observatory of International Migration (OBMigra) reports that Haitian immigrants are considered the largest nationality in the Brazilian labor market. The number of these immigrants in formal employment grew by 50.9% between 2011 and 2013. According to 2014 data from OBMigra, the Haitian population in Brazil grew approximately 18-fold from just over 814 immigrants in 2011 to 14,579 employed in the formal labor market in 2013.

The influx of Haitian foreigners is associated with the changes that have taken place in Brazil's industrial and service sectors, following their development. and services sectors in Brazil, accompanying their development, which has generated great demand for labor, especially those with a less qualified profile (OBMigra, 2014). During this period , many Brazilian companies also came to Acre to select foreigners to work in their companies in other states, as the Haitians' physical willingness to work and their desire and interest in learning were widely seen.

Another situation that occurred was that some services were ignored by Brazilians themselves because they considered the wages to be low, so foreign labor was more than welcome by the companies. For Haitians, the amount was also considered low, especially with the devaluation of the real against the dollar, which reduced the amount they could send to their families in Haiti by converting it to American currency, but Haitians said they were satisfied with the job opportunity.

It is also important to highlight the working conditions that immigrants are subjected to in our country, especially when it comes to the capitalist system which, in order to maintain itself, most of the time increases the working hours, reduces the wages and labor rights of these immigrants, who end up being seen by businessmen as just another potential profit generator for the company.

Businesses have seen them, especially Haitians, as an opportunity to reduce their production costs. Research carried out by British economist Paul Collier for the United Nations (UN) showed that, in 2009, Haiti had a large surplus of skilled labor. According to the study, Haitian workers cost less than Chinese workers (SANCHES, 2014).

In the city of São Paulo, the Catholic Church, through a project that welcomes immigrants called the Mission of Peace, received the immigrants who arrived in Brazil and thus helped them to be included in the labor market. According to this project, between February and April 2012 it is possible to understand how Haitians were being incorporated into the labor market.

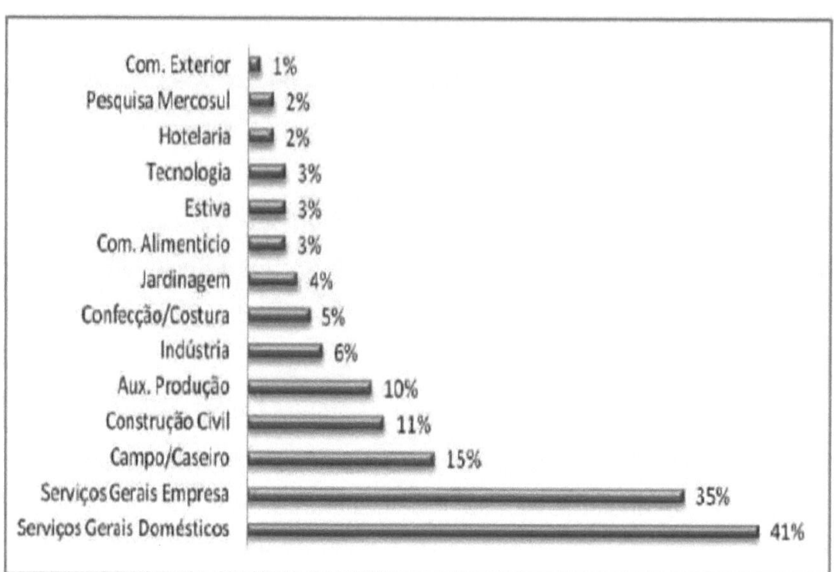

Graph 5: Percentage distribution by branch of activity of those looking for workers in the period from 6/2/2012 to 24/4/2012. Source: Peace Mission Mediation Program, Caffeu, Cutti 2012

This data presented by the peace mission shows an important tool for assessing the occupation of Haitian workers in Brazil. The survey was carried out with the participation of 274 Haitians during the period mentioned above. However, in addition to the research carried out by the peace mission, we have the work coordinated by Fernandes and Castro (2014), also in São Paulo, which provides us with new information (Graph 6).

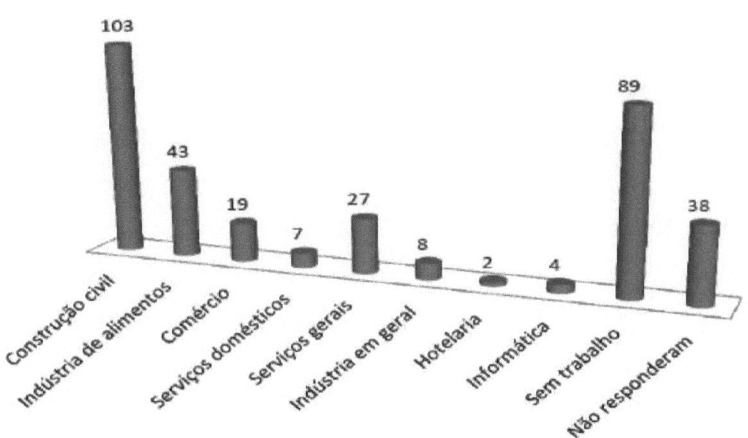

Graph 6: Occupation sector of Haitians in 2013 Source: Fernandes, Castro (2014)

From the interviews conducted by Fernandes and Castro (2014) with 340 Haitians, this seems to be the closest to the reality we have seen in the media regarding the activities carried out by Haitians. The fact that the majority work in construction is due to the fact that in São Paulo it is very well known, which generates great opportunities in the capital and in the metropolitan regions and in the interior of the state. The food industry, which comes in second place, is predominantly related to companies coming from the south of the country, which have found in immigrant workers the opportunity to supplement their workforce due to the lack of labor in their regions.

Of the 2,162 immigrants received by the pastoral this year, 1,363 (52%) have already been employed, according to the priest. Some have been hired by companies in the south, mainly meatpacking plants. Others have been employed in the interior of São Paulo, mainly in construction companies and mechanical workshops (PALHARES, 2014).

Sociology professor Leticia Mamed, from the Federal University of Acre (UFAC), gave an interview to the website *"acaointegrada.org"* and reported on the cases of dentists, doctors, journalists and engineers of various nationalities

43

who are working in construction, heavy industry and chicken and meat slaughterhouses.

The Annual Social Information Report (RAIS) collected information on jobs in Brazil and revealed that the increase in the number of jobs has been occurring in categories with low salaries, such as the administrative sector, commerce and services, civil construction and agriculture, which are characterized by intense turnover in the workforce. Faced with this, in recent years they have increased their hiring of immigrants such as Haitians.

Federative Unklades	2011	2012	2012/2011	2013	2013/2012
Total	79.578	94.688	19,0%	120.056	27,8%
São Paulo	27 515	33.172	20.6%	38 293	15,4%
Rio de Janeiro	9408	11.022	17.2%	11 964	8.5%
Paranâ	2 697	3 390	44.2%	6 544	68.2%
Santa Catarina	1 147	1 875	63.5%	4 376	133.4%
Rio Grande do Sul	1 420	2 181	53,6%	3097	42,0%
Amazonas	1 749	2 089	19,4%	2.225	6.5%
Federal District	1 295	1520	17,4%	1846	21.4%
Minas Gerais	1 245	1 511	21.4%	1827	20.9%
Mato Grosso	712	892	25.3%	1 573	76.3%
Babia	740	845	14.2%	951	12.5%
Rondonia	543	722	33,0%	936	29.6%
Other Federal Units	31.107	34 969	12.4%	46424	32.8%

Table 3: Foreigners with formal employment contracts by state. Source: RAIS/TEM (2011 - 2013)

According to data from the General Register of Employed and Unemployed (CAGED), together with the Ministry of Labor and Employment (MTE), data was obtained on the main Brazilian states where Haitian workers were hired in the formal labor market in 2014.

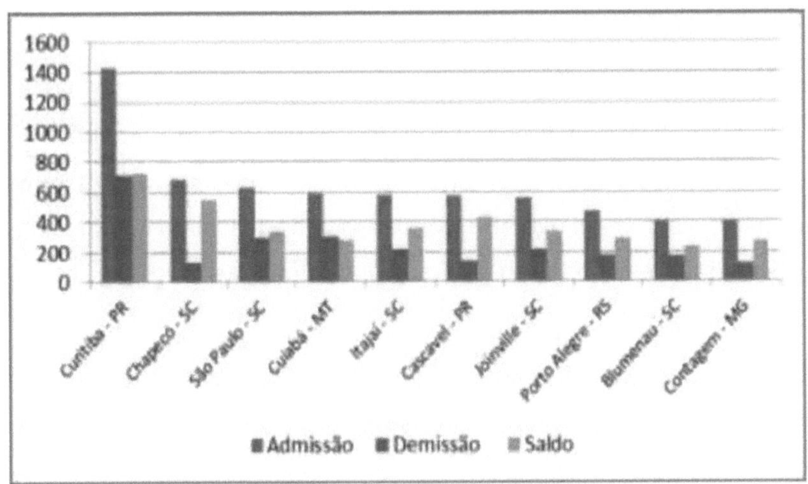

Graph 7: Main Brazilian municipalities according to the number of Haitian workers hired in the formal labor market (2014). Source: CAGED/MTE (2016)

Brazil, through its companies and public authorities, has been incorporating Haitian labor in the way that capitalist power dictates. The desperation caused by the lack of opportunities for the majority of Haitian immigrants who arrive in Brazil in search of new possibilities ends up allowing many companies in sectors such as construction and the food industry to exploit this workforce with longer working hours, non-payment of overtime, a precarious working environment and without essential safety in the workplace.

All this shows us the urgent need to reflect on public policies that really cater for this growing number of immigrants who see Brazil as an opportunity for growth and a change of life.

FINAL CONSIDERATIONS

Brazil's economic growth and major infrastructure projects such as the 2014 World Cup, the 2016 Olympic Games in Rio de Janeiro and the prospect of exploring for oil in the pre-salt layer have turned our country into a new destination for migrants from all over the world, which has resulted in exposing the fragility of national institutions in dealing with situations involving illegal migration.

Migration has been a constant action over the last few centuries in many countries, not just here in Brazil, and this problem can be caused by political or climatic factors, health issues or a lack of social structures. When these migrants go to another country, they need ways to survive and work is seen as a source of survival in the place where they believe they have the greatest opportunities. Thus, the desire of many Haitians to come to Brazil was motivated by the expectation of a better quality of life through work, so that they could send financial resources to help their families survive.

When we begin to understand the historical-geographical formation of Haiti, we realize how the country has become one of the poorest countries in the world and consequently understand the main motivations that lead Haitians to leave their country in search of a decent place to live. Haiti has become a country totally violated by its rulers, full of coups d'état and with the earthquake of 2010, housing conditions, health and job opportunities have only become even more difficult, especially when we look at the history of previous natural disasters that are always occurring.

However, when we look at the situation of these migrants until they arrive in Brazil and the reality they encounter and experience, we realize that it is full of vulnerable situations, which is why there is a real need to go deeper into the analysis of the migratory phenomenon in order to make it possible to build effective and humane policies for dealing with Haitian immigrants. What we currently see is the absence of the state in solving problems related to immigration and only the emergence of the Catholic Church and NGOs as the main subjects in the intermediation between Haitians and incorporation into Brazilian society, which is a serious problem.

Another important conclusion is that there are mechanisms for exploiting the Haitian workforce in several states, from the recruitment process to the working conditions. These include lower salaries than other workers, poor housing and even the use of the Portuguese language as a strategy to violate labor rights in the case of those who don't speak Portuguese well.

We don't deny that people make choices in a migratory process, but these choices are always conditioned in general to social circumstances that go beyond the individual's will and so they must adapt to the conditions immediately encountered. People can make choices, but they are limited by various historical and social events which mean that their wills do not necessarily meet their desires, but rather those situations that reality forces them to accept.

REFERENCES

AGUIAR, A. G. **Five years on, the earthquake that devastated Haiti still leaves its mark on the country**. Agencia Brasil. January 12, 2015. Available at: http://agenciabrasil.ebc.com.br/internacional/noticia/2015-01/cinco-anos-depois-terremoto-que-devastou-haiti-ainda-deixa-marcas-pelo. Accessed on: 24/03/2017

BAENINGER, R. **O Brasil na rota das migraçôes latino-americanas**. Nùcleo de Estudos de Populaçâo-Nepo/Unicamp; Fapesp; CNPq; Unfpa. Campinas. 2013.

BELCHIOR, D. **Haitians are the problem or the problem is the lack of a migration policy that respects immigrants**. Note from the Center for Human Rights and Citizenship of Immigrants and Grito dos Excluidos Continental. In: Carta Capital, São Paulo, April 26, 2014. Available at:

http://negrobelchior.cartacapital.com.br/os-haitianos-sao-o-problema-ou-o-problema- and-the-lack-of-a-migration-policy-that-respects-immigrants/ Accessed on: 17/03/2017.

BRAZIL. **Ministry of Justice**. Available at: http://portal.mj.gov.br. Accessed on: March 25, 2017.

BRAZIL, portal. **Record formal job creation in 2010**. May 11, 2011. Available at: http://www.brasil.gov.br/economia-e-emprego/2011/05/pais-registro-recorde-de-trabalhadores-formais-em-2010-mostra-rais. Accessed on: 01/06/2017.

BRAZIL, Portal. **Job creation hits record in 2010**. Brasil.Gov. January 18, 2011. Available at: http://www.brasil.gov.br/economia-e-emprego/2011/01/apesar-das-407-5-mil-vagas-fechadas-em-dezembro-2010- termina-com-2-5-milhoes-de-novos-empregos. Accessed on: 05/04/2017

COSTA, G. A. **Haitian immigration in Manaus. Presence of the pastoral care of migrants**. In: Travessia - Migrant magazine. N°68, p. 83-89. 2011

COTINGUIBA, G. C. **Haitian immigration to Brazil: the relationship between work and migratory processes**. Dissertation (Master's in History and Cultural Studies) - Federal University of Rondônia/UNIR/RO, 2014.

DE LA BLACHE, P.V **Principios da Geografia Humana**. Lisbon: Cosmos Publishing House, 1954.

FARIA. A. V. de. **The Haitian diaspora to Brazil: the new migratory flow (2010-2012)**. 2012. 136f. Thesis (Master's in Geography). Pontifical Catholic University of Minas Gerais,

2012.

FERNANDES D; CASTRO M. C. (coordinators). **Project "Studies on Haitian migration to Brazil and bilateral dialogue".** Belo Horizonte. 2014.

FERNANDES, D. et. al. **Studies on Haitian migration to Brazil and bilateral dialogue.** Belo Horizonte, MG, 2014. (Study Project. Ministry of Labour and Employment/International Organization Migration - IOM/PUC Minas/Space Distribution of Population Study Group).

FRANCISCO, W. C. **"The Earthquake in Haiti".** Brazil School. Available at <http://brasilescola.uol.com.br/geografia/o-terremoto-no-haiti.htm>. Accessed on 11/03/2017.

FRESNILLO, I. "**Haiti, Four Years After the Earthquake: The Mirage of Reconstruction**", Forum Magazine. January 16, 2014. Available at: http://www.revistaforum.com.br/digital/130/haiti-quatro-anos-depois-terremoto- miragem-da-reconstrucao/. Accessed: 29/03/2017.

GIRALDI, R. **Earthquake toll in Haiti: 220,000 dead and 1.5 million homeless.** Brasilia: Agência Available Available at: <http://agenciabrasil.ebc.com.br/noticia/2011-01-12/saldo-do-terremoto-no-haiti-e-de- 220-thousand-dead-and-15-million-displaced>. Accessed on April 21, 2017.

GOMES, J. **Haitians in Brazil - 1,600 received visas to work and study in the country.** Brasilia: Palmares Cultural Foundation. Available at: <http://www.palmares.gov.br/?p=17191>. Accessed on: April 22, 2017.

HAITI. Available at: <http://www.ebc.com.br/noticias/retrospectiva 2012/2012/12/retrospectiva-imigracao/>. Accessed on: 01/06/2017

HAITI. Available at: http://sosdoaquecimento.blogspot.com.br/2012/11/paises- perifericos-haiti-renda.html>. Accessed on: 10/06/2017.

HAITI. Available at: https://pt.tradingeconomics.com/haiti/indicators>. Accessed on: 14/06/2017.

HAITI. Available at: https://www.terra.com.br/vida-e-estilo/turismo/cruzeiros/ilha- no-haiti-e-destaque-no-route-dos- cruzeiros,e77e49cc33724410VgnVCM20000099cceb0aRCRD.html>. Accessed on: 10/06/2017.

IBGE. BRAZILIAN Institute of Geography AND STATISTICS. **Paisesat**. 2017. Available at: http://www.ibge.gov.br/paisesat/main_frameset. php Accessed on: April 17, 2017.

KLEIN, H. S. **African Slavery, Latin America and the Caribbean**. Translated by José Eduardo de Mendonça. Sao Paulo: Brasiliense, 1987.

MACHADO, A. **Haitians are the majority of immigrants in the formal labor market in Brazil**. Açao Integrada. July 03, 2015. Available at: http://www.acaointegrada.org/haitianos-sao-maioria-entre-imigrantes-no-mercado- de-trabalho-formal-no-brasil/. Accessed on: 06/04/2017

MILESI, R. **International Migration in Brazil**. Contemporary Reality and Challenges. Rio de Janeiro. 2010

MINISTRY OF LABOR. **National Immigration Council**. 2011. Available at: <http://portal.mte.gov.br/cni/>. Accessed on: 21/04/2017.

MANTOVANI F; VELASCO C. **In 10 years, number of immigrants increases 160% in Brazil, says PF**. June 25, 2016. Available at: http://g1.globo.com/mundo/noticia/2016/06/em-10-anos-numero-de-imigrantes- aumenta-160-no-brasil-diz-pf.html. Accessed on: 08/04/2017.

MOREL, M. **The Haitian Revolution and slave-owning Brazil: What should not be said.** - 1 ed. - Jundiai, SP: Paco, 2017.

UNITED NATIONS, **Five years after the earthquake that destroyed Haiti, the UN continues to support the country's reconstruction**. January 12, 2015. Available at: https://nacoesunidas.org/exclusivo-cinco-anos-depois-do-terremoto-que-destruiu-o- haiti-onu-continues-supporting-reconstruction-of-the-country/. Accessed: 31/03/2017

OIMGRA, International Organization for Migration. **Foreign Migration.**

Available at: <http://www.iom.int/jahia/Jahia/about-migration/developing-migrationpolicy/migration-labour/labour>. Accessed on: 02/06/2017.

PALHARES, I. **Haitians spread to companies in the interior of São Paulo State**. Folha de Sao Paulo newspaper. August 18, 2014. Available at: http://www1.folha.uol.com.br/cotidiano/ribeiraopreto/2014/08/1501884-haitianos-se-espalham-pelo-interior-de-sao-paulo.shtml. Accessed on: 04/06/2017.

PEREIRA, A. M. **Haiti - The suffering life of Haitian women**. Brasileiras Pelo Mundo. March 08, 2015. Available at: http://www.brasileiraspelomundo.com/haiti-a- vida-sofrida-da-mulher-haitiana-181612816. Accessed on: 07/04/2017

FEDERAL POLICE. **Foreigners Registration Division** - DICRE. Brasilia, DF, 2014.

PROACRE, **Program for Social Inclusion and Sustainable Economic Development in the State of Acre.** Rio Branco, AC, 2012.

Report on human rights violations in a shelter for Haitians in the city of Brasiléia, northern Brazil. Sao Paulo: Conectas Human Rights, 2013. Available at: <http://www.conectas.org/>. Accessed on: 10/04/2013.

RIBEIRO, C. **The Haitian immigrant in Brazil.** Proceedings of the 6th ABEP Meeting. Sao Paulo. 2015.

SANCHES, M. **Sao Paulo has pilgrimages of businessmen to hire immigrants.** O Globo newspaper. August 17, 2014. Available at: http://oglobo.globo.com/brasil/sao-paulo-tem-romaria-de-empresarios-para-contratar- imigrantes-13633389. Accessed on: 04/06/2017

SILVA, S. A. **Here begins BrazilH. Haitians in the Triple Frontier and Manaus.** Migrations in the Pan-Amazon: flows, borders and socio-cultural processes. Manaus: Fapeam, 2012.

BIBLIOGRAPHY

BARBOSA, V. **Five years after the earthquake that devastated Haiti.** January 12, 2015. Available at: http://exame.abril.com.br/mundo/5-anos-apos-o-terremoto- que-devastou-o-haiti-em-imagens/. Accessed on: 29/03/2017.

BRAZIL. Decree-Law No. 406, of May 4, 1938. **Provides for the entry of foreigners into the national territory.** Chamber of Deputies, Brasilia - DF, Available at: http://2.camara.leg.br/legin/fed/declei/1930-1939/decreto-lei-406-4-   maio-   1938-348724-publicacaooriginal-1-pe.html. Accessed on: 08/04/2017.

CAMARGOS, D. PRATES, M. C. **Brazil is hope for Haitians.** In: Estado de Minas. Belo Horizonte,   March   13,   2011.   National   Editor.   Available   at: http://clipping.radiobras.gov.br/clipping/novo. Accessed on: 26/04/2017.

CARVALHO, C. **Acre suffers from the invasion of immigrants from Haiti.** In: O Globo, Sâo

Paulo, January 1, 2012, Pais section. Available at: http://oglobo.globo.com/brasil/acre-sofre-com-invasao-de-imigrantes-do-haiti- 3549381. Accessed on: 15/04/2017.

CAVALCANTI, L. OLIVEIRA, A. T. TONHATI, T. (Orgs) **A Inserçâo dos Imigrantes no Mercado de Trabalho Brasileiro.** Brasilia: Cadernos do Observatório das Migraçôes Internacionais, 2014.

FERNANDES, D. et. al. **Studies on Haitian migration to Brazil and bilateral dialogue**. Belo Horizonte, MG, 2014 (Study Project. Ministry of Labour and Employment/International Organization Migration - IOM/PUC Minas/Group for the Study of the Spatial Distribution of the Population).

Photos show Haiti's difficult reconstruction five years after the earthquake. G1 Globo. 12 January 2015. Available at: http://g1.globo.com/mundo/noticia/2015/01/fotos-mostram-dificil-reconstrucao-do- haiti-5-years-after-earthquake.html. Accessed on: 20/04/2017

GOMBATA, M. **Haiti: four years after the earthquake, nothing has changed.** August 10, 2014. Available at: https://www.cartacapital.com.br/internacional/reconstrucao- inexistente-deixa-haiti-em-limbo-5533.html. Accessed on: 28/03/2017.

LAW NO. 9.474, OF 22 JULY JULY 22, 1997. - Defines mechanisms for the implementation of the 1951 Refugee Statute, and determines other measures.

MARCEL, Y. NATANI, R. **Acre decrees social emergency due to immigration outbreak.** In: Portal G1, Acre, April 9, 2013. Available at: http://g1.globo.com/ac/acre/noticia/2013/04/acre-decreta-situacao-de- emergenciasocial-por-causa-de-surto-de-imigracao. Accessed on: 02/04/2017.

MARX, K. **Capital**, Volume I - Trad. J. Teixeira Martins e Vital Moreira, Centelha - Coimbra, 1974.

Migrants seek refugee benefits. Zero Hora, Porto Alegre, June 12, 2011. Available at: <http://clipping.radiobras.gov.br/clipping/novo. Accessed on: 16/05/2017.

Brazil: a country of immigration? Electronic Journal of Urban and Regional Studies. Electronic journal @metropolis. N. 09. Year 03. Jun. 2012. P.06-18. Available at: http://www.emetropolis.net/index.php?option=com_edicoes&task=artigos&id=31&lang=en>. Accessed on: 26/03/2017.

ORTIZ, F. **Limit on entry into Brazil is to protect Haitians, says government.** January 29, 2012. Available at: http://operamundi.uol.com.br/conteudo/reportagens/19494/limite+of+entry+into+Brazil+and+for+the+protection+of+Haitians+affirms+government.shtml. Acessoem : 03/04/2017.

PORTAL. **Brazil authorizes permanent residence for 43,800 Haitians**. 11 de October, 2015. Available at: http://www.brasil.gov.br/cidadania-ejustica/ 2015/11/brasil-authorizes-permanent-residence-visa-for-43-8-thousand-haitians. Accessed on: 12/04/2017.

ROCHA, L. ARANHA, A. **O que fazer com os imigrantes do Haiti?** Istoé Magazine. Sao Paulo, February 5, 2011. Brazil section. Available at: http://clipping.radiobras.gov.br/clipping/novo. Accessed on: 09/04/2017.

SEITENFUS, R. **Haiti: international dilemmas and failures**. Ijui: Ed.Unijui, 2014 (Coleçao Relaçôes Internacionais e globalizaçao; 47).

STOCHERO, T. **Daily influx of Haitians triples and the situation is worrying, says Acre government**. Portal G1, Sao Paulo, January 15, 2014. Available at: http://g1.globo.com/ac/acre/noticia/2014/01/em-7-dias-entrada-de-haitianos-triplica- e-acre-temetragedia. Accessed on: 17/03/2017.

TREZZI, H. **Tide of Haitians arrives in Brazil**. In: Zero Hora, Porto Alegre, June 12, 2011. Available at: http://clipping.radiobras.gov.br/clipping/novo. Accessed on: 05/04/2017.

VALLER FILHO, W. **Brazil and the Haitian crisis: technical cooperation as an instrument of solidarity and diplomatic action**. Brasilia, 2007.

VENTURA, D. **Migrating is a right**. Sao Paulo: SESC. February 2, 2016. Available at http://www.sescsp.org.br/online/artigo/s/dFi#/tagcloud=lista. Accessed on 05/04/2017.

ZERO HORA. **Brazil and the Haitians**. In: Zero Hora, Porto Alegre, January 15, 2012. Editorial. Available at: http://wp.clicrbs.com.br/editor/2012/01/12/editorial-apoia- controle-do-ingresso-de-haitianos-no-pais-voce- concorda/?topo=13,1,1,13???0000000. Accessed on: 02/05/2017.

MIX
Papier aus verantwortungsvollen Quellen
Paper from responsible sources
FSC® C105338
www.fsc.org

Printed by Books on Demand GmbH, Norderstedt / Germany